Spiritual DNA

Bridging Science and Spirituality to Live Your
Best Life

Dan Desmarques

22 Lions

Spiritual DNA: Bridging Science and Spirituality to Live Your Best Life

Written by Dan Desmarques

Contents

Introduction

In a world where the pursuit of material success often leaves us feeling empty and disconnected, Spiritual DNA offers a groundbreaking approach to understanding our true nature and unlocking our full potential. This revolutionary book seamlessly weaves together ancient wisdom, cutting-edge science, and practical insights to reveal the hidden blueprint of our existence-our Spiritual DNA.

Drawing from a vast tapestry of knowledge spanning millennia, from the ancient civilizations of Mesopotamia and Egypt to the latest discoveries in quantum physics and neuroscience, this book presents a comprehensive and inclusive perspective that bridges the gap between religion, science, and spirituality.

Discover how your spiritual DNA, much like your biological DNA, encodes the essence of your being, shapes your experiences, and guides your path through the earthly realm. Learn how to:

1. Unlock the mysteries of life and death and access divine wisdom.

2. Navigate life's challenges with greater clarity and purpose.

3. Tap into your innate potential for personal growth and spiritual evolution.

4. Understand the cyclical nature of existence and the lessons we're meant to learn in each lifetime.

5. Align with the fundamental laws of the universe to manifest your dreams and desires.

"Spiritual DNA" doesn't just offer theoretical concepts - it offers specific, pragmatic methods that can be easily verified and experienced by anyone willing to embark on this transformative journey of self-discovery.

As you delve into the pages of this book, you'll discover

- The scientific basis for the interconnectedness of all things, from subatomic particles to cosmic consciousness.

- How to harness the power of your thoughts and emotions to create your reality.

- The role of spiritual DNA in expanding human consciousness and potentially facilitating the evolution of our biological DNA.

- Practical techniques for accessing higher states of consciousness and tapping into your innate wisdom.

- The profound implications of aligning with your spiritual DNA for personal relationships, career success, and overall well-being.

This book is not just a read-it is an invitation to transform your life and contribute to a collective shift in human consciousness. By recognizing and aligning with your spiritual DNA, you'll open the door to unprecedented personal growth, deeper connections with others, and a more harmonious relationship with the natural world.

Whether you're a spiritual seeker, a science enthusiast, or simply someone who wants to live your best life, Spiritual DNA offers a unique and powerful framework for understanding your place in the universe and realizing your full potential. It's time to unlock the blueprint of your existence and step into the life you were meant to live.

Embark on this transformative journey today and discover the incredible power that lies within your spiritual DNA. Your best life is waiting for you - are you ready to claim it?

Chapter 1: The Essence of Life and Death

T hroughout the ages, humanity has sought to understand the profound mysteries of existence, grappling with questions of life, death, and the nature of consciousness. From the ancient civilizations of Mesopotamia, Babylon, and Egypt to the philosophical traditions of Europe, a vast tapestry of knowledge has been woven, offering insights to promote better lifestyles and holistic health. In recent times, however, we seem to have strayed from this path of wisdom and become entangled in emotional turmoil, physical pain, and a myriad of diseases.

Ancient explorations of the spiritual dimensions of human existence assert that spirituality is both the alpha and the omega of every life cycle, serving as the fundamental framework for understanding the meaning of life and the purpose of each individual on earth. These diverse sources present a comprehensive and inclusive perspective that bridges the gap between religion, science, and superstition.

Central to this discourse is the concept of Spiritual DNA, a unique and personal imprint of our spiritual journey across multiple lifetimes. Much like our biological DNA, this spiritual counterpart encodes the essence of our being, shaping our experiences and guiding our path through the earthly realm. Spiritual DNA serves as the key to unlocking the mysteries of life and death, providing a gateway to paradise and the wisdom of the divine.

This concept transcends the boundaries of any single religious belief, scientific paradigm, or social construct. It stands as a supreme truth manifested throughout human history in every expression of intellect and creativity. The current state of societal degeneration, marked by violence, mental instability, and widespread disease, can be attributed to our collective disconnection from this vital source of life.

Spiritual DNA plays a critical role in the evolution of our consciousness, potentially facilitating the expansion of our biological DNA beyond its current double helix structure. It is the catalyst for enlightenment and the source of wisdom. A thorough understanding of this concept is not merely academic; it is essential to the future of humanity and our liberation as spiritual beings.

Historically, this knowledge has been wielded as a tool of power, used to establish religions with millions of followers or to form secret societies of influential individuals. But by weaving together these diverse strands of thought, we reveal the interconnectedness of all spiritual and scientific endeavors, demonstrating that each perspective, while valuable, is incomplete without acknowledging its relationship to the greater whole.

This approach eschews esoteric or mystical obfuscation and instead offers specific and pragmatic methods that can be readily verified and experienced by anyone willing to embark on this journey of self-discovery. The universality of this immortal truth lies in its accessibility and resonance with the human experience across cultures and time.

As we delve into the nature of Spiritual DNA, we uncover its role in shaping our individual and collective destiny. This concept provides a framework for understanding the cyclical nature of existence, the purpose of our earthly incarnation, and the lessons we are meant to learn in each lifetime. By recognizing and aligning with our spiritual DNA, we can navigate life's challenges with greater clarity and purpose, fostering personal growth and spiritual evolution.

The implications of this understanding go far beyond individual enlightenment. As more and more people awaken to the reality of their Spiritual DNA, a collective shift in consciousness becomes possible. This awakening has the potential to transform society by fostering a greater sense of interconnectedness, compassion, and harmony with the natural world.

In addition, the concept of Spiritual DNA offers a bridge between the material and spiritual realms, providing a framework for integrating scientific discoveries with spiritual insights. As quantum physics continues to reveal the fundamental interconnectedness of all things, the ancient wisdom encoded in our Spiritual DNA finds validation in the language of modern science.

Chapter 2: The Essence of Human Existence and Consciousness

T hroughout human history, the quest to understand the nature of our existence and consciousness has been a central theme in philosophy, religion, and science. Recent advances in various fields, including psychology, neuroscience, and quantum physics, have shed new light on the intricate relationship between our physical bodies, our consciousness, and what many refer to as the spiritual realm. This exploration has led to the concept of Spiritual DNA, a profound and multifaceted idea that seeks to explain the nature of human existence and our connection to divine or universal consciousness.

The concept of Spiritual DNA is based on the understanding that human beings are more than their physical bodies. Each individual possesses a unique energetic signature, often referred to as an aura, which can be perceived by some and measured with specialized equipment. This aura is not static; it fluctuates based on various factors, including our physical health, emotional state, and spiritual well-being. The aura serves as an external manifestation of our inner self, emanating from the core of our being and reflecting the state of our soul.

In this context, soul can be understood as our manifesting spiritual identity, distinct from but connected to our Source or Spirit. Spirit, in turn, is connected to what many describe as the Field of Unlimited Energy, Infinite Intelligence, or God - the ultimate matrix of life. This hierarchical structure of spirit, soul, and personality allows us to navigate the complexities of existence on multiple levels of reality.

Our spirit, as the most fundamental aspect of our being, has the ability to manifest in and adapt to any reality or planetary system. The soul represents the specific path we've chosen to experience in this lifetime, while our personality is a complex interplay between our spiritual essence, societal expectations, and the experiences we accumulate in our current incarnation.

The concept of spiritual DNA suggests that we carry with us the accumulated wisdom and skills from our past lives that are imprinted in our spiritual essence. When we live in alignment with our spiritual nature, we can access this vast store of knowledge and experience, leading to personal growth, enhanced intuition, and a deeper understanding of our purpose in life. Conversely, when we become overly attached to material pursuits and disconnected from our spiritual essence, we can experience a range of physical and psychological ailments, including degenerative diseases, anxiety, and stress.

This understanding is consistent with various spiritual traditions and modern scientific research. For example, Jose Silva's mind experiments demonstrated that the brain operates more efficiently at lower frequencies, a state he called the alpha level. At this level, Silva postulated that individuals could connect more deeply with the spiritual dimension, gain insight into their purpose in life, and access higher forms of intelligence.

Similarly, many religious texts, including the Bible, speak of humanity's divine nature and our potential to embody godlike qualities. The concept of being "created in the image of God" (Genesis 1:27) and the call to be "like God in true righteousness and holiness" (Ephesians 4:24) suggest a profound connection between human consciousness and divine consciousness.

The spiritual purpose of life, viewed through the lens of Spiritual DNA, is to learn how to transform reality with consciousness, manifest our dreams with purpose, and experience the results with responsibility. This perspective encourages us to actively engage in our spiritual growth, recognizing that our thoughts, beliefs, and actions have far-reaching consequences not only in our immediate reality, but also in the broader cosmic tapestry.

Recent studies in the field of epigenetics have shown that our experiences and environment can influence gene expression, lending scientific credibility to the idea that our consciousness can influence our physical reality. This interplay between the spiritual and material realms suggests that our spiritual DNA is not just a metaphysical concept, but may have tangible effects on our biological processes.

In addition, the concept of spiritual DNA provides a framework for understanding phenomena such as spontaneous spiritual awakenings and kundalini experiences. These profoundly transformative events, characterized by a sudden sense of union with a perceived ultimate reality, can catalyze drastic shifts in perception, worldview, and well-being. Research has shown that these experiences, while sometimes challenging at first, are overwhelmingly perceived as positive and can lead to long-term improvements in mental health and spiritual well-being.

Chapter 3: The Interplay of Spirituality and Human Existence

The Buddha's teachings, while profound in their recognition of the spiritual realm, have often been misinterpreted by followers who promote a philosophy of non-action as action. This misinterpretation fails to capture the essence of Buddha's message, which was not about inertia, but rather about mindful presence.

Buddha's famous words, "Do not dwell on the past and do not dream of the future, but focus the mind on the present moment," represent a philosophy of cognition, not a complete philosophy of life. It's important to understand that Buddhism, like yoga, is not meant to be an all-encompassing philosophy of life, but rather a tool for spiritual growth and self-knowledge.

The danger lies in confusing these spiritual practices with comprehensive philosophies of life. If we were to accept the Taoist notion that action and non-action are the same, we would be questioning the very purpose of our existence on earth. This perspective fails to take into account the multifaceted nature of human experience and the diverse realities that exist throughout the universe.

To truly embrace our spirituality and understand our place in the cosmos, we must move away from religious dogma and recognize the limitations of human interpretation. The truth of God and our spiritual nature cannot be fully grasped through philosophies of non-action or oversimplified maxims such as "the contented are rich." Such statements, while potentially mistranslated or distorted over time, fail to capture the limitless nature of Truth and its ability to manifest in myriad forms.

When we consider the vastness of the universe, with its countless worlds and beings existing in realities both denser and more ethereal than our own, we begin to see the incompleteness of these ideas. Our life on Earth is not a mistake, but rather a purposeful stage in our spiritual evolution. Arrogance corrupts the mind and invites the influence of a lower spiritual nature, while humility opens us to a higher understanding.

The fact that we cannot manifest everything we desire or fully conquer our subconscious but primitive drive to compete and envy suggests that we are still on earth for a reason. However, philosophies such as Buddhism and Taoism often fall short in understanding this reason by dismissing the importance of physical action. We cannot avoid action altogether, for non-action contradicts the essential laws of our reality.

As we explore the spectrum of existence, from worlds of greater suffering to realms of heightened magic and freedom, we see that the concepts of freedom and power take on different meanings. On Earth, freedom may be associated with travel and economic power, while in higher realms it may mean unconditional love and unlimited artistic creation. Similarly, power on Earth is often associated with control, while in the higher realms it's more closely tied to consciousness and understanding.

The essence of God-like consciousness is to humbly accept our current state of being while seeking divine wisdom through the elements of reality presented to us. These elements include time, energy, and space, which we perceive through our senses. As we expand our perception, we gain a richer understanding of our reality and our place within it.

When individuals with similar insights or dreams connect, they create a void that opens up a parallel reality. What was once a shared fantasy becomes a shared reality, creating a new state of existential purpose. This transformation is rooted in the realization that there are no coincidences in human interactions - we are all destined to meet, though recognizing this truth is a choice we must make.

Hidden perceptions exist in every aspect of our lives, from our actions to our relationships. By becoming aware of these perceptions, we unlock our karma, initiate rapid life changes, and allow intuition from past lives to flow naturally into our consciousness. This process of self-discovery leads us to our most vital truth - our spiritual DNA.

To truly understand why we are not yet as God intended, we must expand our expression and grow in self-awareness. In doing so, we become more God-like, melting away the illusions of the material world through an elevated consciousness that empowers and transforms our very essence.

Chapter 4: The Interconnectedness of Mind, Body, and Self-Determination

The intricate relationship between our mind, body, and self-determination forms the foundation of our existence and shapes our journey through life. This complex interplay redefines and reshapes our soul, allowing for unlimited development and progress throughout our lives. Our spiritual essence is not separate from our physical form, but is systematically connected to it and manifested through our capacity for self-determination.

The illusion of free will in the material world serves as a guide to consciousness, leading us to experience what we need rather than what we merely want. Paradoxically, on a spiritual level, we tend to desire what we truly need and experience what is necessary for our growth. This apparent contradiction highlights the multifaceted nature of our existence, which encompasses both the physical and spiritual realms.

In the manifestation of spirituality, there are no failures, only experiences. Suffering arises from the need to experience limitation, which itself serves as a pillar of unknown truth. The full perception of truth can only be achieved through the experience of untruth, which gives meaning to our lives. Life itself can be conceptualized as energy moving between different frequencies, resulting in an infinite expression of creativity in manifestation. This can be visualized as an electric field with a specific spectrum of waves or strings interacting with each other.

Within this spectrum, the earth and human frequencies are of paramount importance. These frequencies interact in a transcending effort to evolve into more sublime forms of manifestation, moving from the heaviest to the lightest, with light itself being the highest form. Scientific research has identified various brain wave patterns that correspond to different states of consciousness, ranging from beta waves (associated with intense alertness) to delta waves (associated with deep sleep or relaxation).

The levels of brainwave activity correspond to different states of being and cognitive functions. In the highest state, we experience an absence of thought, with awareness based on intuition. The middle state involves thought directed toward creativity, visualization, and dreaming. The lowest state is characterized by constant thinking, often influenced by environmental factors.

Researcher Jose Silva's experiments have shown that lower cycles per second of brain activity correlate with a quieter mind, which facilitates connection with what he calls "higher intelligence." This suggests that we can gather information from both the conscious state (the material world) and the subconscious or deeper level of consciousness (the spiritual world or God's Source). By choosing to connect to both realms with a clear state of mind, we can better understand and fulfill our purpose on Earth.

The brain can be seen in this context as a tool of the conscious mind, useful for interacting with the physical world, but potentially becoming a barrier when attempting to connect with the spiritual realm. The expansion of our spiritual DNA depends on our ability to avoid being trapped by uncontrolled thoughts and emotional stimuli such as fear, worry, anxiety, and anger.

Achieving an ideal balance involves accepting our emotions beyond the illusion of reality, rather than suppressing them. This allows us to understand a higher realm beyond the perceptions our brain can assimilate, leading to the expansion of the soul in consciousness. Knowledge plays a crucial role in this process, feeding our soul with the seeds of perception. However, the most valuable information is that which speaks of Truth, a concept that has often been corrupted throughout history as unprepared souls are not meant to ascend spiritually and access these revelations prematurely.

The supreme revelation lies within each individual and can be felt when living a spiritual life. The material world can only present assumptions while denying the truth, for it lacks the decoding spirit necessary to truly understand its implications. Unprepared minds may even fear the truth because of its potential impact on their lives, which explains why prophets throughout history have often been persecuted or dismissed.

As Jesus Christ is quoted in the Gospel of Thomas, the kingdom of God is both within and without us. Self-knowledge is the key to realizing our true nature as "sons of the living Father. Without this self-knowledge we remain in a state of spiritual poverty.

The world is full of coincidences, but few truly perceive their deeper meaning. Only through a careful study of history can we begin to unravel the seeds of life, the composition of our spiritual DNA, and its purpose. However, even historical records are often lacking in truth and discernment.

Chapter 5: The Interconnectedness of Humanity and Nature

I n the tapestry of existence, humanity and nature are inextricably woven together, forming a complex and beautiful pattern that reflects the essence of life itself. This profound interconnectedness, long recognized by indigenous cultures and increasingly supported by scientific research, offers a compelling framework for understanding our place in the world and the path to spiritual enlightenment.

The wisdom of Native American traditions emphasizes the fundamental unity between humans and the natural world. "Whatever we do to the web, we do to ourselves; all things are connected," Chief Seattle wisely observed in 1854. This perspective is not just poetic sentiment, but a profound ecological truth that modern science is beginning to confirm.

Recent studies in the field of biophysics, such as those conducted by Baker (1995), have provided fascinating insights into the vibrational frequencies of trees. These studies have found that healthy trees have frequencies between 0.2 Hz and 2 Hz, which, interestingly, are the lowest frequencies a human being can reach. This correlation suggests a deep resonance between human consciousness and the natural world, and the possibility of achieving harmony by aligning ourselves with the rhythms of nature.

The destruction of forests can therefore be seen not only as an ecological crisis, but also as a spiritual one. Each tree lost represents a diminution of the world's capacity for balance and sanity. As Standing Bear poignantly stated, "The heart of man becomes hard when he is separated from nature." This hardening is not only metaphorical, but can have tangible effects on our mental and spiritual well-being.

The concept of Mother Earth, revered in many indigenous cultures, takes on new meaning in light of these scientific findings. The words of Big Thunder Bedagi Wabanaki Algonquin resonate deeply: "The Great Spirit is in all things, the Great Spirit is our Father, but the Earth is our Mother. She nourishes us..." This nourishment seems to extend beyond the physical into the vibrational and energetic realms.

Interestingly, studies of animal brain wave frequencies provide further evidence of our connection to the natural world. Electroencephalography (EEG) research has shown that adult humans typically exhibit brain wave frequencies of 14 Hz and higher, while various animals exhibit lower frequencies: rats at 6-10 Hz, monkeys at 7-9 Hz, and cats and rabbits at 4-6 Hz. This hierarchy of frequencies challenges our conventional notions of intelligence and superiority, suggesting that higher frequencies do not necessarily equate to greater wisdom or harmony with the natural world.

Indeed, the human condition often seems to be characterized by a paradoxical disconnection from nature, despite our technological advances. We exploit natural resources for material gain and selfish needs, often at the expense of the very systems that sustain us. This short-sighted approach not only jeopardizes our own survival, but also diminishes the potential for future generations to thrive on this planet.

The consequences of this disconnect manifest themselves in myriad ways. Children born in today's world face unprecedented challenges: exposure to genetically modified organisms, polluted air and water, and high levels of electromagnetic radiation. These environmental stressors can alter not only our physical DNA, but also our spiritual essence. The rising incidence of learning disabilities, depression, and other mental health problems could be seen as symptoms of a damaged "spiritual DNA" - our innate capacity for connection, empathy, and transcendence.

Ancient wisdom traditions, such as those found in the Upanishads, offer guidance for reconnecting with our spiritual essence. "Respect food, for food sustains the body, and the body exists to serve the soul" (The Upanishads 4:10). This simple yet profound teaching reminds us of the sacred nature of our physical existence and its role in supporting our spiritual journey.

However, it's important not to fall into the trap of oversimplification or dogmatism. As the Upanishads warn, "Religious people who devote themselves to rituals and sacrifices are ignorant of their ignorance" (5:5). True spirituality transcends rigid adherence to ritual or denial of our unique human capacities. Instead, it calls us to embrace our full potential as conscious, creative beings capable of co-creation with the divine.

Socrates eloquently captured this aspiration: "The end of life is to be like God, and the soul that follows God will be like Him." This perspective invites us to see our human abilities not as a source of superiority over nature, but as a sacred responsibility to care for and nurture the world around us.

Our spiritual DNA can then be understood as our innate capacity for cooperation, creativity, and compassion in service to life on Earth. It's our responsibility to evolve and develop the world, to learn from our actions, and to create a reality that promotes the purest frequencies of Earth and supports life in all its diverse manifestations.

This path of spiritual evolution and ecological harmony requires a fundamental shift in our values and priorities. As the Bible warns in 1 Timothy 6:10, "The love of money is the root of all evil." The pursuit of material wealth at the expense of ecological and spiritual well-being has brought us to the brink of crisis. This doesn't mean that all wealth or progress is inherently destructive. As the Bhagavad Gita teaches, "Selfless action is inspired by God"

(5:16). When our actions and innovations are motivated by a desire to serve and uplift all of life, they become channels for divine expression.

Jesus' teachings echo this sentiment, promising that those who give up worldly attachments for the sake of higher ideals "will certainly receive much more in this life and have eternal life in the world to come" (Luke 18). Similarly, the Vedas affirm that "the wealth of those who give generously is never exhausted" (2:1). These spiritual truths point to a model of abundance that comes from generosity, cooperation, and alignment with the laws of nature.

Imagine a world in which our technological prowess is used not for exploitation but for the nurturing of life in all its forms. We have the knowledge and ability to create clean energy systems, automated processes that free humans from drudgery, and computer systems that support rather than supplant natural ecosystems. Such a world would offer unprecedented opportunities for creative expression, spiritual growth, and the fulfillment of our deepest dreams and aspirations.

The obstacles to realizing this vision lie not in our lack of ability, but in our collective mindset. Our self-destructive tendencies, rooted in fear, greed, and a sense of separation from nature, prevent us from fully utilizing the resources and knowledge already at our disposal. By healing this fundamental disconnect and embracing our role as conscious stewards of the earth, we can begin to manifest a reality that more closely resembles heaven on earth.

Chapter 6: The Complex Dance of Human Frequencies

As we move through life, our brainwave patterns evolve to reflect our growth and development. This fascinating progression begins with the delta waves of infancy, moves through the theta waves of early childhood, blossoms into the alpha waves of adolescence, and finally settles into the beta waves of adulthood. This natural progression of brainwave frequencies offers a profound insight into the human psyche and our interactions with the world around us.

Consider the innocence and wonder of a child whose brain waves resonate at frequencies that allow for unbridled imagination and openness to new experiences. As Socrates wisely observed, "An honest man is always a child." This observation speaks to the purity and authenticity that children embody, qualities that many adults strive to recapture in their quest for personal growth and self-actualization. Those who claim to dislike children may actually be revealing a disconnection from their own inner child, having drifted too far from the frequencies of youth.

Our brainwave frequencies not only reflect our individual stages of development, but also influence our social interactions and preferences. We tend to gravitate toward those who resonate at similar frequencies, finding comfort and understanding in their presence. This phenomenon explains why we often feel most comfortable with peers of similar age or life experience. However, it's important to recognize that these frequencies are not fixed and can be influenced by our environment, our experiences, and our conscious efforts to expand our awareness.

As we navigate the complex web of human relationships, we encounter individuals across a spectrum of frequencies. In the highest and most positive ranges we find those who radiate a nurturing energy that promotes the survival and well-being of those around them. These individuals often express a deep love for nature, animals and children, and embody a harmonious connection with the world around them. Their presence can be uplifting and inspiring, encouraging others to tap into their own positive frequencies.

Conversely, those who vibrate at lower, more destructive frequencies may be drawn to expressions of anger, fear and domination. This attraction to negative energies can manifest in many ways, from a fascination with violent media to a desire for power and control over others. It's important to recognize these tendencies in ourselves and others, as they can have far-reaching consequences on both a personal and societal level.

The effects of frequency resonance extend beyond individual interactions to shape entire environments. Sensitive individuals often develop an intuitive ability to read the energy of a room or situation, sometimes experiencing premonitions or gut feelings about future events. This heightened awareness can be both a gift and a challenge, requiring careful navigation of various social and environmental contexts.

Our relationship to nature serves as a powerful indicator of our frequency alignment. Those who resonate at higher, more positive frequencies often find solace and rejuvenation in natural environments, feeling a deep connection to the earth and its rhythms. In contrast, individuals vibrating at lower frequencies may feel uncomfortable or disconnected in nature, preferring the artificial stimulation of urban environments or technology-driven experiences.

The modern world presents unique challenges to maintaining positive frequency alignment. The proliferation of electromagnetic frequencies from Wi-Fi, mobile devices and other electronic sources can potentially disrupt our natural brainwave patterns. This is especially evident in densely populated urban areas, where individuals can be bombarded with a cacophony of artificial frequencies. It's important to be aware of these influences and to take steps to protect and nurture our natural rhythms.

Interestingly, certain professions seem to attract people with certain frequency tendencies. For example, the high suicide rates among mental health professionals, physicians, and law enforcement officers may reflect the intense emotional and energetic demands of these roles. These statistics underscore the importance of maintaining energetic balance and seeking support when navigating challenging professional environments.

As we reflect on these findings, we're reminded of the wisdom of Aristotle: "A friend to all is a friend to none". This statement speaks to the importance of discernment in our relationships and the value of authentic connections. Those who resonate on frequencies similar to our own often become our closest confidants and supporters, while maintaining healthy boundaries with those on very different frequencies can be critical to our well-being.

Chapter 7: The Hidden Dimensions of Spiritual Growth

T he journey of spiritual growth and enlightenment is a profound and challenging one that has been pursued by seekers throughout human history. As we delve into the depths of spiritual consciousness, we encounter a complex interplay of forces that shape our experiences and relationships with the world around us. This exploration reveals fundamental truths about the nature of reality, consciousness, and our place in the cosmos.

At the heart of spiritual development is the realization that we are more than our physical bodies and material existence. Ancient wisdom traditions and modern scientific insights converge in the understanding that consciousness and energy are the fundamental building blocks of reality. As the Bhagavad Gita eloquently states, "All visible things arise from the invisible." This profound truth echoes through the ages and finds resonance in the words of visionaries like Nikola Tesla, who marveled at the electrical design of our planet and its potential to enable miracles.

The pursuit of spiritual growth often sets individuals apart from mainstream society as they begin to resonate on different frequencies. This divergence can lead to feelings of isolation and rejection, as described in biblical passages that warn against attachment to worldly desires. However, this separation is a natural consequence of evolving consciousness. As Jose Silva insightfully observed, "On Beta, we hunt each other; on Alpha, we pray for each other." This shift in frequency and perspective fundamentally changes the way we interact with others and perceive the world.

The concept of frequency and resonance extends beyond the metaphysical realm into the very fabric of physical reality. Dr. Hideki Yukawa's groundbreaking work on the strong nuclear force revealed that even at the subatomic level, particles interact through the exchange of virtual particles, creating the bonds that hold matter together. This scientific understanding parallels spiritual teachings about the interconnectedness of all things and the unseen forces that shape our reality.

As we progress on our spiritual journey, we may find ourselves opening up to higher dimensions of consciousness. The idea of evolving beyond the third dimension into fourth, fifth, and higher realms of existence speaks to the limitless potential for growth and expansion of consciousness. This evolution is not just an individual journey, but part of a collective shift, as the Earth itself is believed to be resonating at higher frequencies. This planetary evolution offers the potential for a more uplifting level of existence, provided we can overcome our destructive tendencies and embrace a more harmonious relationship with our environment.

The Law of Attraction emerges as a powerful principle in this spiritual framework, suggesting that our thoughts and energies play a crucial role in shaping our reality. This concept is consistent with quantum physics theories about the observer effect and the role of consciousness in collapsing wave functions into tangible reality. By understanding and harnessing this principle, we can become more conscious co-creators of our experiences.

The cyclical nature of existence, as expressed through concepts such as reincarnation, offers a broader perspective on our spiritual evolution. Every life experience, potentially across multiple worlds and dimensions, offers opportunities for growth and learning. This expansive view challenges our linear perception of time and space, echoing Einstein's assertion that the distinction between past, present, and future is merely a "stubbornly persistent illusion."

Embracing spiritual growth requires courage and resilience, as it often means stepping away from conventional paths and facing the unknown. The rejection and misunderstanding one may face from others can be seen as a testament to one's progress rather than a deterrent. As we cultivate our "spiritual DNA," we may find ourselves increasingly out of sync with those who remain focused solely on material existence. However, this divergence also opens doors to deeper connections with like-minded individuals who share our frequency and aspirations.

The eyes, often referred to as the windows to the soul, play a significant role in spiritual perception and connection. The ability to perceive and connect with others on a soul level transcends verbal communication and allows for instant recognition of kindred spirits or discordant energies. This intuitive knowing guides us in forming relationships and navigating social interactions from a place of higher consciousness.

Ultimately, the path of spiritual growth is one of continuous evolution and expansion. As Aleister Crowley so eloquently stated, "The joy of life is in the exercise of one's energies, in constant growth, in constant change, in the joy of each new experience. To stop is simply to die". This perspective encourages us to embrace the challenges and opportunities for growth that life presents, recognizing them as essential steps on our spiritual journey.

Chapter 8: Resonance, Choice, and Self-Concept

I n the complex tapestry of human relationships, we often find ourselves drawn to individuals who reflect both our deepest needs and our most profound rejections. Far from being a mere coincidence, this phenomenon is deeply rooted in the principles of attraction and the quantum nature of our reality. Those versed in the Law of Attraction understand that in order to manifest a soul mate, one must embody and radiate the very qualities and emotions one seeks in a partner. This process is not just about wishful thinking, but about aligning one's entire being with the desired outcome.

The concept of choice plays a central role in this dance of attraction. As Kevin Michel eloquently states, "Choice is what presents us with a multitude of paths, because choice creates a flow of electrons through the brain in a way that leads inexorably to quantum superposition and the many worlds that are the inevitable result." This profound insight suggests that our choices, even at the subatomic level, shape the reality we experience. Our subconscious mind, aware of these myriad possibilities, selects the reality we inhabit based on our self-concept - the deeply held beliefs and perceptions we have about ourselves.

Consciously or unconsciously, we tend to attract individuals who resonate on the same frequency as we do. This resonance is not always apparent on the surface, but operates at a deeper, often subconscious level. Those who dream of finding someone who will fundamentally change them or their lives often find themselves either alone or in relationships that feel incomplete or imperfect. This search for an idealized other to "fix" us is a misguided attempt to outsource our personal growth and happiness.

Relationships, like our bodies, are inherently perfect systems. It is our unhealthy habits and misconceptions that introduce imperfections and challenges. The common belief that true love is an unattainable ideal doesn't reflect the nature of love itself, but rather the frequency we radiate. To attract the partner we desire, we must first align ourselves with the frequency of that desired reality. This alignment involves more than wishing or hoping; it requires a profound shift in our sense of self and the energy we project into the world.

The power of conscious desire leads us to dream, which leads us into a state of creative illusion. It is within this illusion that we must believe in order to manifest a new reality. The line between what we consider a dream and what we perceive as reality is blurred, much like the distinction between potential reality and optional reality. Our spiritual DNA contains the blueprint for all that we can become and experience. As Bob Proctor insightfully states, "Science and psychology have isolated the single most important cause of success or failure in life. It is the hidden self-image you have of yourself."

This self-image is critical to the attraction process. Just as a dog cannot attract another dog by acting like a cat or a bird, we cannot attract experiences or relationships that are fundamentally misaligned with our true nature and self-image. We always attract what we need rather than what we consciously want because our deepest needs shape our desires, often in ways we don't fully understand.

The paradox of human desire often leads us to want what we unconsciously reject, creating a cycle of unfulfillment. For example, a person who is fundamentally selfish may desire love but have difficulty attracting it because their energy resonates with selfishness rather than acceptance and giving. Such a person is likely to attract similarly selfish partners, resulting in a frustrating cycle of mutual recrimination and disappointment.

At its core, the desire to possess is itself a selfish impulse. True love transcends desire; it is an act of giving without expectation or need. The art of loving, then, can be distilled to the practice of selfless giving. Regardless of our actions or identity, if we cannot give without attachment, we cannot truly love. Love is the awareness of giving without possessing, a state of being rather than a transaction.

Before we can hope to attract a loving partner, we must first cultivate self-love and radiate that energy into the world. It's impossible to radiate a frequency of self-love deficiency and expect to attract someone operating on a different wavelength, able to give us what we cannot give ourselves. This principle underscores the importance of personal growth and self-acceptance on the path to meaningful relationships.

Chapter 9: The Power of Sound in the Human Experience

E motions and frequencies intertwine in a complex dance, creating a symphony that resonates throughout our being. This fascinating interplay of sound, emotion, and consciousness has captivated scientists, philosophers, and spiritual practitioners for centuries. As we delve deeper into this realm, we uncover profound insights that challenge our understanding of reality and offer new pathways to healing, transformation, and spiritual growth.

At the heart of this exploration is the concept that emotions have frequency, and frequency has sound. This idea suggests that our emotional states are not just abstract experiences, but are rooted in the physical world of vibration and energy. When we encounter certain sounds or music, they can trigger specific feelings and evoke powerful memories, suggesting a direct link between auditory stimuli and our emotional landscape.

The Six Solfeggio Frequencies, a set of ancient musical tones, have received considerable attention in recent years for their purported healing properties. These frequencies, each associated with specific effects on the human psyche and physiology, offer a tantalizing glimpse into the potential of sound as a therapeutic tool. From releasing guilt and fear (396 Hz) to facilitating spiritual awakening (852 Hz), these tones are believed to resonate with different aspects of our being, promoting balance and well-being.

This concept of frequency-based healing is not limited to the Solfeggio scale. Researchers have studied how different types of music and sounds can affect our emotional states and even our physical health. For example, uplifting music with higher frequencies can increase our mood and energy levels, while soothing, low-frequency sounds can induce relaxation and meditative states. The ancient syllable "Om," revered in many Eastern spiritual traditions, is said to cancel out the force of gravity when intoned at certain low frequencies, suggesting a profound connection between sound, consciousness, and the physical world.

The power of sound to shape our reality goes beyond mere emotional responses. Max Planck, the father of quantum theory, postulated that all matter arises from and exists because of a force that causes particles to vibrate. This force, he suggested, must be guided by a conscious and intelligent mind-a matrix underlying all matter. This perspective is consistent with ancient wisdom traditions that view the universe as a vast, interconnected web of vibrations, with sound playing a crucial role in its creation and maintenance.

Nikola Tesla, another visionary scientist, emphasized the universal responsiveness of all matter to external stimuli, challenging the notion of inert, "dead" matter. He saw ignorance as the greatest obstacle to human progress and advocated the spread of knowledge and the unification of humanity as the way forward. This holistic view of reality, in which even seemingly inanimate objects possess some form of consciousness, resonates with both cutting-edge scientific theories and ancient spiritual teachings.

The implications of these ideas are profound and far-reaching. If our emotions and thoughts are indeed frequencies that interact with the physical world, then we have the power to shape our reality by consciously manipulating these vibrations. Practices such as meditation, visualization, and sound healing take on new meaning in this light, offering tools for personal transformation and spiritual growth.

In addition, this perspective challenges us to rethink our relationship with the natural world. If all matter is imbued with consciousness and responds to vibrations, then our interactions with the environment take on a new dimension of responsibility and potential for harmony. The Vedic concept of a divine syllable flowing through all creation echoes this interconnectedness, suggesting that by attuning ourselves to the frequencies of nature, we can align ourselves with a higher cosmic order.

As we explore these fascinating concepts, however, it's important to maintain a balanced perspective. While mathematical laws are powerful tools for understanding reality, they may not capture its full essence. The relationship between our models of reality and reality itself is complex and often elusive. This humility in the face of the mysteries of the universe is essential as we continue to explore the boundaries of science and spirituality.

In practical terms, this understanding of the relationship between sound, emotion and consciousness opens new avenues for personal growth and healing. By consciously choosing the sounds and frequencies we expose ourselves to, we can potentially influence our emotional states, enhance our well-being, and even facilitate spiritual awakening. Techniques such as sound baths, binaural beats, and frequency-based meditation offer promising tools for those who wish to explore these possibilities.

As we continue to explore the emotional and frequency landscape of human experience, we are called to approach this knowledge with both curiosity and discernment. The potential for sound and frequency-based interventions to improve human life is immense, but it must be pursued with scientific rigor and ethical consideration. By integrating ancient wisdom with modern scientific understanding, we can open new pathways to healing, personal growth, and collective harmony.

Chapter 10: Reconnecting with Nature's Frequency for Optimal Health

In our modern world, we have become increasingly disconnected from the natural rhythms and energies of the earth. This disconnection, largely driven by technological advances and urbanization, has had a profound impact on our health and well-being. Recent scientific research is shedding light on the importance of reconnecting with the earth's frequency, a practice known as "earthing" or "grounding," which may hold the key to addressing a wide range of chronic health issues plaguing our society.

The concept of grounding is rooted in the understanding that the Earth possesses a negative electrical potential generated by solar winds, the ionosphere, and lightning storms. For most of human history, our ancestors maintained direct skin contact with the Earth's surface, allowing a continuous flow of electrons between their bodies and the planet. This natural connection served to support various body systems and scavenge free radicals that promote inflammation.

However, the advent of modern lifestyles, particularly in the post-World War II era, has dramatically reduced our physical contact with the earth. The widespread use of insulating materials in shoes, for example, has effectively disconnected us from the Earth's energy field. This disconnection coincides with a significant shift in the leading causes of death in the 21st century, with chronic degenerative diseases overtaking infectious diseases as the leading health concern.

Dr. Stephen Sinatra, founder of the Heart MD Institute, offers a compelling analogy to illustrate the impact of this disruption on our health. He suggests that healthy blood should flow smoothly, like red wine, but many people today have thicker blood, more like ketchup. This increased viscosity puts more strain on the heart and increases pressure in the arteries and blood vessels, contributing to various cardiovascular problems.

Research has shown that pain in the body is often associated with an electron deficiency. In response to this deficiency, blood cells tend to clump together in an attempt to acquire the electrical charge they need. Unfortunately, this clumping triggers an inflammatory, free-radical response that perpetuates the cycle of pain and inflammation.

The practice of grounding provides a natural solution to these problems by allowing the free electrons of the earth to enter the body. This influx of electrons helps synchronize our bioelectrical systems and significantly reduces inflammation. The more compromised one's health is, the greater the potential benefits of grounding.

Clint Ober, a retired cable television executive, drew a fascinating parallel between the human body and cable television signals. He observed that grounding cables to earth virtually eliminates signal interference, and that all electrical systems are stabilized by grounding. Applying this principle to the human body, Ober proposed that grounding is a "universal regulating factor in nature" with profound effects on bioelectrical, bioenergetic, and biochemical processes.

The simplicity and accessibility of grounding make it an attractive therapeutic approach. Experts suggest that as little as 15 minutes of direct skin contact with the earth's surface-whether by walking barefoot on the ground or touching a tree rooted in the earth-can have significant benefits. This practice has been shown to reduce the effects of radiation on the body by nearly 100%, even when simultaneously exposed to radioactive sources such as cell phones or computers.

In addition to its protective effects against electromagnetic fields (EMFs), grounding has been associated with a wide range of health benefits. These include improvements in allergies, asthma, arthritis, autism, chronic sinusitis, cancer, depression, dementia, heart disease, obesity, osteoporosis, and various autoimmune diseases. In addition, grounding has been linked to increased energy and vitality, reduced inflammation and oxidation, enhanced intelligence, regulated blood pressure, improved skin conditions, and increased longevity.

The therapeutic potential of grounding extends beyond direct physical contact with the earth's surface. Ancient civilizations recognized the healing properties of clay, a substance intimately connected to the earth's frequency. The first recorded use of healing clay dates back to ancient Egypt and Mesopotamia around 2500 BC, where it was used as an anti-inflammatory and antiseptic. The Ebers Papyrus from 1500 B.C. describes the use of clay for a variety of ailments, including intestinal and eye problems.

Modern research has confirmed the benefits of clay for various health conditions. Clays have been shown to effectively remove excess oils and toxins from the skin, making them valuable in the treatment of dermatological problems such as boils, acne, ulcers, abscesses, and seborrhea. In addition, recent studies suggest that clay consumption may aid in the removal of radioactive particles from the body. The negatively charged clay particles attract positively charged toxins in the stomach, preventing their absorption through the intestinal walls into the bloodstream.

The healing power of earth-related substances and practices can be understood through the lens of frequency. All elements of life emit a certain frequency, and those that are connected to the earth's frequency support life by strengthening and protecting our DNA. This understanding bridges the gap between our biological and spiritual selves, as Carl Sagan articulated: "The cosmos is within us. We are made of star stuff. We are a way for the universe to know itself."

As we grapple with the challenges of modern life, including the pervasive influence of technology and our increasing disconnection from nature, the practice of grounding offers a simple yet powerful means of rebalancing our bodies and minds. By reconnecting with the earth's natural frequency, we can improve intercellular communication, reduce inflammation, improve overall system function, and ultimately achieve optimal health.

Chapter 11: Nutrition, Health, and Longevity

A s we delve deeper into the relationship between food and well-being, a compelling narrative emerges-one that blends ancient wisdom with modern scientific understanding and challenges our current approach to nutrition and health. The foundation of this narrative is a simple yet profound concept: the closer our food is to the earth's natural frequencies, the more it contributes to our overall health. This idea is not just New Age philosophy, but is supported by anthropological data. Across cultures and throughout history, humans have shown a preference for nutrient-dense foods, especially animal products and fats. However, it's important to note that these traditional diets were fundamentally different from what we consider "animal products" today (Kumar, 2018; Sikalidis, 2018).

In ancestral diets, foods were free of modern additives, preservatives, and artificial colors. They lacked added sugars and refined flours and were not subjected to industrial processing methods. Dairy products were consumed in their natural, unaltered state - not pasteurized, homogenized, or reduced in fat content. Both plant and animal foods were grown in pesticide-free soil and were not treated with growth hormones or antibiotics. In essence, our ancestors consumed what we would now call "organic" foods (Kumar, 2018).

This stark contrast with modern dietary patterns raises important questions about the impact of our current food system on health and longevity. In today's industrialized world, it's becoming increasingly rare for individuals to reach the century mark. This phenomenon coincides with a food landscape dominated by genetically modified organisms (GMOs) and highly processed products (Kumar, 2018; Mehboob, 2023).

The situation is further complicated by the complex relationship between food industry giants, pharmaceutical companies, and health insurance companies. For example, large health insurance companies in North America have significant investments in fast food companies. At the same time, the pharmaceutical industry operates a multi-billion dollar enterprise that some critics argue is more focused on profit than health (Kumar, 2018).

This complex web of interests has led to a food environment in which we are constantly exposed to a variety of chemicals, often with little regard for their long-term health effects. These substances have been linked to several chronic diseases, including stroke and cancer, many of which are thought to be caused by disruptions at the DNA level. In fact, it's estimated that over 95% of all chronic diseases can be attributed to dietary choices, toxic food ingredients, nutritional deficiencies, and physical inactivity (Kumar, 2018; Mehboob, 2023; Perera et al., 2021).

The consequences of this shift away from traditional whole-food diets are evident in populations that have rapidly transitioned to Western-style diets. For example, Native Hawaiians, who once thrived on a diet rich in coconuts, fish, shellfish, taro, sweet potatoes, and fresh fruits, now face alarming rates of obesity and diabetes after adopting a more processed, Western-style diet (Kumar, 2018; Sikalidis, 2018).

However, the impact of diet on health goes beyond physical ailments. It's estimated that approximately 70% of all physical illnesses have a psychosomatic component, suggesting that our mental and emotional states play a significant role in our overall health. This holistic view of health is consistent with the teachings of ancient healers such as Hippocrates, who famously stated, "The natural forces within us are the true healers of disease" (Cole, 2014; Kumar, 2018).

This perspective shifts the role of health professionals from merely treating symptoms to facilitating the body's innate ability to heal. As Thomas Edison prophetically stated, the physician of the future should "cure and prevent disease with nutrition" (Kumar, 2018).

Indeed, numerous studies have demonstrated the potential of plant-based diets to prevent and even reverse chronic diseases such as cancer, diabetes, and hypertension. These findings echo the ancient wisdom of Hippocrates: "Let food be your medicine and medicine be your food" (Bye et al., 2021; Huang et al., 2022; Kumar, 2018).

Certain nutrients have shown particular promise in supporting health and fighting disease. For example, vitamin B1 (thiamine), found in beans, rice, vegetables, and citrus fruits, has been used in anti-radiation therapies to detoxify the body. Similarly, iodine, which is abundant in seaweed and seafood, has been linked to lower cancer rates in populations with high dietary intake, such as the Japanese (Kumar, 2018; Yang et al., 2023).

The concept of food energy, while often overlooked in conventional nutritional science, provides another level of understanding of the relationship between diet and health. This perspective suggests that the energetic quality of food-influenced by factors such as how it was grown or raised-can impact our well-being beyond its nutritional content (Kumar, 2018; Rathore et al., 2023).

As we navigate the complexities of modern diets, it's clear that a return to whole, minimally processed foods is critical for optimal health. However, this shift must be accompanied by a broader reassessment of our food system, including how food is produced, processed, and marketed. It also requires a more holistic approach to health that considers not only what we eat, but also how we live, think, and interact with our environment (Dixon et al., 2023; Lawrence, 2022).

Chapter 12: Advances in Brain-Computer Interfaces

The intersection of neuroscience and technology has given rise to a new frontier in human enhancement and medical treatment: brain-computer interfaces (BCIs). This rapidly evolving field traces its roots back to the early 20th century, when visionaries such as Nikola Tesla first explored the therapeutic potential of electromagnetic frequencies. Today, we are on the cusp of a revolution that promises to transform our understanding of the brain and our ability to interact with the world around us.

The journey from Tesla's experiments to modern BCIs is a testament to human ingenuity and scientific progress. Tesla's work with specific frequencies to relieve pain and improve blood flow laid the groundwork for electrotherapy and diathermy treatments. These early forays into manipulating the body's electrical systems hinted at the vast potential of bioelectric intervention. As our understanding of neuroscience and technology has advanced, so has our ability to interface directly with the brain.

The development of Radio Frequency Identification (RFID) technology marks a significant milestone in this journey. Initially used for mundane purposes such as identifying VIP customers in nightclubs, RFID chips quickly found applications in healthcare. Hospitals began using them to track patients and staff, and soon the technology was being implanted in patients to store medical information. This evolution from external tags to implantable devices foreshadowed the more invasive brain implants that would follow.

The potential applications of brain implants are staggering. Researchers and pharmaceutical companies envision a future in which neurological disorders such as epilepsy, Parkinson's disease, and even depression can be treated by direct electrical stimulation of the brain. Early experiments in rats have shown promising results in restoring motor function to damaged brain tissue, offering hope to stroke victims and those suffering from neurodegenerative diseases. But the implications of this technology go far beyond medical treatments.

The ability to interface directly with the brain opens up possibilities that were once the realm of science fiction. Imagine downloading knowledge directly into your brain, controlling electronic devices with your thoughts, or communicating telepathically with others. These scenarios are no longer far-fetched fantasies, but potential realities on the horizon.

Companies like Intel and IBM are at the forefront of this research, developing sensors and algorithms that can interpret brain activity and translate it into actionable commands. The gaming and entertainment industries are likely to be early adopters of this technology, but its potential applications are virtually limitless. From increasing productivity in the workplace to revolutionizing the way we interact with our environment, BCIs could fundamentally change the human experience.

Perhaps most intriguing is the potential for direct brain-to-brain communication. Researchers at the University of Washington have already demonstrated a rudimentary form of this technology, allowing one person to control another's hand movements with thoughts alone. As this technology advances, we may see the development of true telepathic communication, allowing the exchange of complex thoughts and emotions without the need for spoken or written language.

However, as with any transformative technology, the rise of BCIs brings with it a host of ethical and societal concerns. The ability to directly influence brain function raises questions about privacy, autonomy, and the nature of human consciousness. Who will have access to this technology and how will it be regulated? What are the implications for personal identity and free will if our thoughts can be read, influenced, or even controlled by external devices?

The potential for abuse is also significant. While the medical applications of BCIs offer hope to millions suffering from neurological disorders, the same technology could potentially be used for nefarious purposes. The ability to manipulate brain function could be exploited for mind control, surveillance, or even as a weapon. As we move further into this brave new world of neurotechnology, it is critical that we proceed with caution and careful consideration of the ethical implications.

The development of BCIs also raises profound philosophical questions about the nature of humanity and our relationship with technology. As we become more integrated with our devices, where does the line between human and machine become blurred? Will augmented humans with BCIs create a new form of inequality, dividing society into the augmented and the unaugmented?

Despite these concerns, the potential benefits of BCI technology are too great to ignore. From restoring function to the disabled to enhancing human cognitive abilities, BCIs represent a new frontier in human evolution. As we continue to unlock the secrets of the brain and develop more sophisticated ways to interface with it, we must strive to balance progress with ethical considerations.

The future of BCIs is both exciting and daunting. As we stand on the brink of this neurotechnological revolution, it is critical that we approach its development and implementation with wisdom, foresight, and a deep respect for the complexity of the human mind. Only by carefully navigating the promises and perils of this technology can we hope to harness its full potential for the betterment of humanity while preserving our essential human qualities.

Chapter 13: The Convergence of Technology and Consciousness

I n an era of rapid technological advancement, we stand on the precipice of a profound transformation of human cognition and experience. The integration of brain-computer interfaces (BCIs) and implantable microchips into the human body has moved from the realm of science fiction to an increasingly tangible reality. With approximately 100,000 people worldwide already living with brain implants, primarily for medical purposes, the U.S. government's commitment to invest over $70 million to advance brain implant technology signals a new frontier in human-machine symbiosis.

This technological leap promises to revolutionize our interaction with information and our perception of reality itself. As envisioned by Larry Page, the future may hold a world where knowledge is instantly accessible by mere thought, facilitated by neural implants that seamlessly interface with our cognitive processes. This vision of frictionless access to information could fundamentally change our relationship to knowledge, learning, and decision-making.

The potential ubiquity of this technology is staggering. Michael Snyder's prediction that brain implants could become as commonplace as smartphones are today underscores the potential for widespread adoption and integration into daily life. This shift could usher in a new era of human capabilities, enhancing our cognitive abilities and expanding our understanding of the world around us in ways previously unimaginable.

Interestingly, this technology may also provide empirical validation for concepts long relegated to the fringes of scientific discourse, such as the Law of Attraction. Intel's research using functional magnetic resonance imaging (fMRI) has shown that certain thought patterns correlate with distinct changes in brain activity, suggesting a tangible link between our thoughts and physiological responses. This finding lends credence to the idea that our mental states can influence our reality, albeit through mechanisms more grounded in neuroscience than metaphysics.

However, the promise of enhanced cognitive capabilities and seamless access to information is not without its downsides. The potential for these technologies to be used as tools of control and manipulation looms large in the minds of many. The ability to interface directly with the brain opens up unprecedented possibilities for influencing thoughts, emotions, and behaviors. In the hands of a tyrannical government or malicious actors, such technology could become the ultimate tool of oppression, potentially programming citizens to experience artificial states of contentment or compliance.

This dystopian scenario raises profound questions about the nature of free will, autonomy, and the nature of human consciousness. If our thoughts and feelings can be externally manipulated or artificially induced, what becomes of our individual identity and agency? The prospect of a "natural high" that never ends, while superficially appealing, threatens to disconnect us from the full spectrum of human experience, including the struggles and challenges that shape our character and drive our personal growth.

Moreover, the integration of such technology into essential aspects of daily life, such as financial transactions, raises concerns about privacy, security, and personal freedom. The biblical prophecy of a future in which commerce is contingent upon bearing a "mark" takes on new meaning in light of these developments, highlighting age-old fears about control and surveillance in a technologically advanced society. But in the midst of these concerns, it's important to recognize the potential benefits of this technology, particularly in medical applications. For people with neurological disorders or physical disabilities, brain implants offer the hope of restored function and improved quality of life. The challenge is to harness these advances for the betterment of humanity while protecting against their misuse.

As we navigate this new frontier, ethical considerations must be at the forefront of technology development and implementation. Transparency, informed consent, and robust safeguards against unauthorized access or manipulation of neural interfaces are paramount. In addition, preserving spaces free from digital intrusion and maintaining the individual's ability to disconnect will be critical to preserving human autonomy and mental well-being.

The integration of technology with human consciousness represents both an extraordinary opportunity and a significant responsibility. It offers the potential to expand our understanding of the mind, enhance our cognitive abilities, and overcome the limitations imposed by physical disabilities. But it also challenges us to address fundamental questions about the nature of consciousness, free will, and what it means to be human in an increasingly digital world.

Chapter 14:
The Architect
of Human-Computer
Interaction

I n the annals of scientific history, few figures loom as large as Nikola Tesla, a man whose visionary ideas and groundbreaking inventions laid the foundation for much of our modern technological landscape. As we stand on the cusp of a new era of human-computer interaction and artificial intelligence, it is both fitting and illuminating to revisit Tesla's prescient insights and examine how they continue to shape our understanding of technology and its potential to transform society.

At the dawn of the 20th century, Tesla envisioned a world of global interconnectedness, a concept that would not be fully realized for another hundred years. His "World System" was a bold reimagining of communication, anticipating the broadcast of music, speech, images, and even navigation signals across the globe. This vision of a unified, interconnected humanity resonates strongly with our current reality, where the Internet and mobile devices have become the backbone of global communication and information sharing.

Tesla's vision went beyond simple communication. He predicted personal communication devices that could fit in a pocket, a concept that seemed outlandish at the time but has become ubiquitous in the form of smartphones. His understanding of frequency and information processing laid the foundation for technologies we now take for granted, from television to Wi-Fi. The ability to transmit complex data - images, sound, text - through the air using specific frequencies is a direct realization of Tesla's theories.

As we move deeper into the 21st century, Tesla's ideas continue to find new applications and interpretations. The field of nanotechnology, for example, is exploring ways to apply similar principles to enhance human cognitive abilities. The concept of neuroimplants that allow direct brain-computer interaction is no longer the stuff of science fiction, but is becoming a tangible reality. This development raises profound questions about the nature of human consciousness and our relationship to technology.

The increasing integration of computers into our daily lives reflects Tesla's vision of a world in which technology and humanity become increasingly intertwined. From tools for work and communication to extensions of our emotional and social selves, computers are becoming indispensable companions in the human condition. This trend points to a future where the lines between human and machine become increasingly blurred, a concept often referred to as transhumanism.

As we move toward this technologically enhanced future, however, we must also grapple with the ethical and societal implications of such advances. The potential for brain-computer interfaces and artificial intelligence to enhance human capabilities is immense, but so are the risks of abuse and the erosion of privacy and autonomy. Tesla's own life serves as a cautionary tale in this regard - despite his genius, many of his most revolutionary ideas were suppressed or co-opted by those with less altruistic motives.

The parallels between computer programming and human cognition offer intriguing possibilities for the future of human-computer interaction. As we become more accustomed to digital interfaces and constant connectivity, the prospect of direct, telepathic communication between humans and machines becomes increasingly plausible. This convergence of biology and technology could lead to unprecedented advances in fields ranging from medicine to education to artistic expression.

In the midst of this technological revolution, however, we must not lose sight of the fundamental aspects of human nature that Tesla himself emphasized. He spoke of instinct that transcends knowledge, of finer fibers that allow us to perceive truths beyond logical deduction. In our rush to improve and expand our capabilities, we must be careful not to suppress these uniquely human qualities.

Tesla's vision of wireless power transmission, if fully realized, could have dramatically altered the course of human development, potentially freeing us from dependence on fossil fuels and centralized power structures. While this particular dream remains unrealized, it serves as a powerful reminder of the transformative potential of visionary thinking and the importance of pursuing ideas that challenge the status quo.

As we continue to push the boundaries of human-computer interaction and artificial intelligence, we would do well to heed Tesla's words about the Earth becoming a "huge brain" capable of responding in every one of its parts. This metaphor takes on new meaning in the age of the Internet of Things and ubiquitous computing. Our challenge is to ensure that this global network enhances rather than diminishes our humanity.

The concept of frequency, central to much of Tesla's work, finds a modern parallel in discussions of consciousness and spirituality. The idea that our thoughts and intentions can influence reality resonates with both quantum physics and ancient wisdom traditions. As we develop technologies that interface more directly with human consciousness, we may gain new insights into the nature of reality and our place within it.

By balancing technological advancement with ethical consideration and respect for human nature, we can work toward a future that realizes the best of Tesla's vision-a world in which technology serves to unite and uplift humanity, rather than divide and control it. In doing so, we may yet fulfill Tesla's dream of a more connected, harmonious, and enlightened global society.

Chapter 15: Revealing the Resonance of Human Needs

Although some individuals have suppressed technologies that could have existed a hundred years ago, and many powerful figures continue to suppress advances in science, medicine, and technology, we must also remember that we participate in our world with our needs and desires, for they resonate at frequencies that shape our reality. This profound insight invites us to explore the depths of our consciousness and the vast expanse of energy that surrounds us. By understanding the interplay between our inner world and the universe at large, we can unlock the secrets of fulfillment and purpose.

Central to this exploration is the concept of resonance. Just as Nikola Tesla demonstrated the power of resonant frequencies by creating an earthquake with a steam-driven oscillator, we too can harness the power of resonance in our lives. Tesla's experiment, conducted nearly 60 years before scientific confirmation, serves as a testament to the untapped potential within us and the natural world. This remarkable feat illustrates how aligning ourselves with the right frequency can produce extraordinary results.

Meditation emerges as a powerful tool in this context. As Jose Silva aptly observed, when our minds are still and our bodies are relaxed, we're still conscious but able to connect with the subconscious. This practice allows us to tune into frequencies that expand our energy field while shielding us from outside influences. It's a gateway to accessing our inner wisdom and connecting with the universal life force.

The idea that we can control nature as much as we can control ourselves stems from the fundamental principle of interconnectedness. Buddha's concept of "being without being" encapsulates this state where human needs cease to exist and we become one with the universe. This oneness with all that is allows us to tap into an infinite source of energy and potential.

Our dreams and aspirations, no matter how seemingly unattainable, are rooted in our spiritual essence. Nikola Tesla's words resonate deeply here: "The gift of mental power comes from God, the Divine Being, and when we focus our minds on this truth, we become in tune with this great power." This perspective invites us to see our desires not as mere material wants, but as expressions of our spiritual DNA and catalysts for higher consciousness.

However, the path to fulfilling our desires is not always straightforward. Before we manifest our dreams, we often go through experiences that challenge us to understand the underlying reasons for our needs. This karmic process serves as a crucial step in our spiritual growth. It's important to realize that obtaining our desires without understanding their deeper meaning can lead to unfulfillment or even destruction.

L. Ron Hubbard's assertion that "your greatest ability is to get an idea" underscores the creative power of thought. Our material reality is shaped by our mental constructs, underscoring the importance of cultivating positive and purposeful thoughts. However, we must be careful not to confuse material illusions with spiritual purpose, lest we become ensnared in the flames of greed and selfishness.

Buddha's wisdom offers guidance in navigating this complex terrain: "A wise person, realizing that the world is only an illusion, does not act as if it were real and thus escapes suffering." This profound insight encourages us to seek truth and happiness as interconnected paths leading to our true destiny. If we fail to recognize this truth, we risk losing touch with our authentic selves and the frequency of salvation.

Courage plays a vital role in this journey of self-discovery and manifestation. As L. Ron Hubbard eloquently stated, courage is "The willingness to cause and to proceed against all odds to achieve the effect one has postulated." It requires that we face our fears, acknowledge our weaknesses and accept the truth of our existence.

In essence, the path to fulfillment and purpose involves a delicate balance between understanding our needs, aligning with universal frequencies, and courageously pursuing our dreams. By recognizing the illusory nature of the material world while purposefully engaging with it, we can transcend suffering and connect with our higher selves.

This journey of self-discovery and manifestation is not about accumulating material possessions or achieving worldly success. Rather, it's about attuning ourselves to the resonant frequencies of the universe, understanding the deeper meaning of our desires, and using them as tools for spiritual growth and higher consciousness.

As we walk this path, we must remain mindful of the power of our thoughts and the responsibility that comes with them. Our ability to manifest our dreams is a testament to our connection to the Divine and the infinite potential within us. However, this power must be wielded with wisdom, compassion, and a deep understanding of our true purpose.

Chapter 16: The Path to Personal Fulfillment

Those who achieve greatness in any field share a common trait: an unwavering belief in their ability to succeed. They refuse to entertain doubt and instead focus their energy on finding solutions and overcoming obstacles. Conversely, those who falter along the way often succumb to a cycle of doubt, seeing more obstacles than opportunities.

This stark contrast in mindset becomes evident when we observe how individuals respond to potential solutions to their problems. While some eagerly embrace new possibilities, others erect mental barriers, inundating themselves with "maybes" and "not sures". As more solutions are presented, these individuals tend to create more excuses, effectively blocking their own path to success. Their attitude becomes a self-fulfilling prophecy, reinforcing their perceived inability to achieve their goals. But part of paying the price is the willingness to do whatever it takes to get the job done. It comes from a declaration that no matter what it takes, no matter how long it takes, no matter what comes up, you will get it done. This unwavering commitment to one's goals is a critical element in overcoming the obstacles that inevitably arise on the path to success.

Interestingly, the act of sharing our goals with others can sometimes have unintended consequences. When we expose our intentions to those who doubt our abilities or harbor jealousy, we inadvertently create a powerful energetic frequency that opposes our goals. This phenomenon explains why some people believe that discussing their dreams with friends brings bad luck. In reality, it is the collective energy of disbelief and negativity that can hinder our progress.

Paradoxically, our closest friends and family members are often the strongest forces against our dreams, even if they are unaware of it. Their fear of losing us or their inability to imagine our success can be a formidable barrier to our aspirations. This is why success depends on getting good at saying no without feeling guilty. You cannot move forward with your own goals if you are always saying yes to someone else's projects. You can only move forward with the lifestyle you want if you focus on the things that will create that lifestyle.

To overcome the negative energy that comes from others, we must cultivate a higher frequency of belief in ourselves. This requires recognizing that rejection is ultimately a construct of our own mind. To overcome rejection, you must realize that rejection is really a myth. It doesn't really exist. It is just a concept that you have in your mind.

The image that others have of us can greatly affect our ability to manifest our desires. People may oppose our goals for a variety of reasons, including fear of becoming less important in our lives, fear of losing us altogether, or the belief that we are undeserving of success. These external perceptions, combined with our self-image, create a complex web of influences that can make personal transformation challenging.

Often, significant life changes require us to risk everything we have, including our relationships and support systems. As L. Ron Hubbard wisely stated, "Never regret the past. Life is in you today, and you make your tomorrow." Those who take the greatest risks and undergo the most profound changes often find themselves alone, but this loneliness can be a catalyst for personal growth and self-discovery.

As some experts have noted, when you're 18, you worry about what everyone thinks of you; when you're 40, you don't give a damn what anyone thinks of you; when you're 60, you realize that no one thought of you at all. This underscores the futility of basing our actions on the opinions of others. People are primarily concerned with their own lives and expectations, which means that unless we live authentically, we risk living someone else's life instead of our own.

Our reality is shaped by the collective frequencies, both positive and negative, of those around us. Every thought resonates at a certain frequency and is connected to the energies of those we know. This intricate web of influences means that we manifest our desires in proportion to what we don't need, the unawareness of others, and the strength of our own beliefs.

Paradoxically, it is often easier to achieve what no one expects. Those who are deeply committed to their goals may lose friends along the way. As Hubbard noted, "Basically, your possibilities are much better than anyone has ever allowed you to believe." This realization can be both liberating and isolating as we free ourselves from the limitations imposed by the expectations of others.

One of the biggest mistakes we can make is relying on others to fulfill our dreams or expecting their approval. This attitude can lead to disappointment and stagnation. Instead, we must accept the possibility that we may have to pursue our goals alone. Hubbard's advice rings true in this situation: "An individual must rise above an ardent desire for the approval of a humanoid group in order to accomplish anything worthwhile."

In our modern society, many individuals are shaped by outside influences, particularly the mass media. Television and other forms of media often serve as conduits for implanted values that control minds and perpetuate societal norms. This creates a dichotomy in life: we can either be our own project or, for fear of losing credibility and social status, become a project of outside forces.

Those with strong self-esteem and self-determination often find themselves at odds with society, facing rejection, job loss, and humiliation. This has been the fate of true visionaries throughout history, individuals with what might be called a strong spiritual DNA.

The strongest frequency we can harness manifests as action, while the weakest manifests as apathy. We are either actively changing the world or slowly dying in it. Our world is constantly being maintained, destroyed and uplifted by various forces. Those who fear change maintain the status quo, while those who profit from destruction actively work against progress. It is the individuals who embrace an uplifting reality and strive for a better world who truly promote positive change.

Paradoxically, our world is often controlled by entities operating at the lowest energy levels, such as pharmaceutical companies and military industries, while those operating at the highest frequencies, such as artists, often struggle to make ends meet. In light of this reality, Hubbard's words resonate deeply: "There is only one way you will ever have a future: make one."

Chapter 17: Aligning Thoughts, Emotions, and Actions

M oney, like every other form of energy in the universe, responds to the laws of vibration. Just as we can attract people or experiences into our lives, we can attract financial abundance. However, this process is often misunderstood or oversimplified. The truth is that attracting wealth requires a holistic approach that includes our thoughts, emotions, beliefs and actions.

At its core, our relationship with money is deeply influenced by our beliefs and emotional state. As Bob Proctor says, "Money is the easiest thing in the world once you learn how to do it. It's like driving a car. It's easy once you know how to do it." This analogy emphasizes that attracting wealth is a skill that can be learned and mastered, but it requires understanding the underlying principles and practicing them consistently.

Our emotions play a crucial role in this process. They create an energy field that attracts experiences and circumstances that match their frequency. This concept is consistent with scientific principles, as Bill Bryson eloquently puts it: "Protons give an atom its identity, electrons its personality." In the context of wealth attraction, our core beliefs about money shape our identity, while our emotions and thoughts shape our financial "personality" and, consequently, our experiences with money.

One of the paradoxes of attraction is that the more desperately we need or want something, the harder it becomes to attract it. This is because the energy of need or lack is at a different frequency than the energy of abundance. John D. Rockefeller, the world's first billionaire, understood this principle when he said, "If your only goal is to get rich, you will never get rich." True wealth attraction comes from aligning with the frequency of abundance rather than focusing on the absence of money.

This alignment requires a shift in consciousness. Instead of dwelling on what we lack, we must cultivate the feeling that we already have what we desire. This explains why people often get what they want when they least expect it or when they no longer feel they need it. It's not about luck; it's about getting into a state of being where you are vibrationally aligned with your desires.

Visionary thinking is a powerful tool in this process. As Bob Proctor noted, "All the great achievers of the past were visionary figures; they were men and women who projected into the future. They thought about what could be rather than what already was, and then they took action to make those things happen." This ability to vividly imagine and emotionally connect with a desired future state is a key component of attracting abundance.

The concept of parallel realities in quantum mechanics offers a fascinating perspective on this process. As Kevin Michel explains, "The more we delve into quantum mechanics, the stranger the world becomes; appreciating this strangeness of the world, while still operating within what you now consider reality, will be the foundation for shifting the current trajectory of your life from ordinary to extraordinary." This suggests that by aligning our thoughts and emotions with our desired reality, we can effectively "shift" into a parallel world where our dreams come true.

However, negative emotions and limiting beliefs often get in the way of this alignment. Those struggling to achieve financial abundance often harbor thoughts such as "I don't deserve it," "I can't do it," or "I will fail," accompanied by feelings of fear, anxiety, and self-doubt. Overcoming these mental and emotional barriers is critical to attracting abundance.

Reprogramming our thoughts and feelings is a powerful strategy. It involves consistently affirming positive beliefs about our worthiness and ability, such as "I can do it" and "I deserve it," while engaging in activities that bring joy and fulfillment. This practice helps shift our vibrational frequency to one that's more aligned with abundance.

Another critical component is inspired action. As Nicola Tesla observed, motion and energy are fundamental to the manifestation of matter. Similarly, our thoughts and emotions must be coupled with consistent, inspired action toward our goals in order to manifest our desires in the physical world.

Belief in one's own deservingness is perhaps the most powerful energy one can emit in the process of attracting abundance. As Bob Proctor says, "Thoughts become things. If you see it in your mind, you will hold it in your hand." This belief must be deeply rooted and supported by a strong spiritual or philosophical foundation.

Many wealthy individuals attribute their success to a higher power and view their ability to create wealth as a divine gift. John D. Rockefeller famously said, "God gave me my money" and "The power to make money is a gift from God to be developed and used to the best of our ability for the benefit of mankind." This perspective not only strengthens their belief in their deservingness, but also imbues their wealth with a sense of purpose and responsibility.

In conclusion, attracting abundance requires shifting our consciousness from a state of scarcity to one of abundance, cultivating visionary thinking, overcoming limiting beliefs, and taking inspired action. By understanding and applying these principles, we can learn to navigate the quantum field of possibility and manifest the financial abundance we desire, not just for personal gain, but as a means of making a positive contribution to the world around us.

Glossary

Alpha Level: A state of brain activity characterized by lower frequencies (8-13 Hz), associated with relaxation, creativity, and increased connection to spiritual dimensions.

Beta Level: A state of brain activity characterized by higher frequencies (14-30 Hz), associated with normal waking consciousness and active thinking.

Brain wave patterns: Different frequencies of electrical activity in the brain that correspond to different states of consciousness and cognitive function.

Divine Consciousness: A state of consciousness aligned with the highest spiritual frequencies, characterized by unconditional love, wisdom, and oneness with all creation.

Earth Realm: The physical dimension of existence, characterized by material experiences and limitations.

Electromagnetic frequencies: Waves of electrical and magnetic energy that can affect human consciousness and well-being, especially in urban environments with high technological density.

Enlightenment: A state of spiritual awakening characterized by profound understanding, inner peace, and connection to higher consciousness.

Frequency Alignment: The process of attuning one's personal energy to specific vibrational frequencies, often associated with spiritual growth and well-being.

Frequency Resonance: The phenomenon whereby individuals or environments vibrate at similar energetic frequencies, influencing thoughts, emotions, and behaviors.

Higher Intelligence: A concept referring to a level of consciousness or wisdom beyond ordinary human perception, often associated with spiritual or divine realms.

Karma: The spiritual principle of cause and effect in which actions and intentions in one lifetime influence experiences in future lifetimes or spiritual growth.

Life Cycle: The continuous process of birth, death, and rebirth over multiple lifetimes, each contributing to spiritual evolution.

Manifestation: The process of bringing thoughts, desires, or intentions into physical reality through focused awareness and alignment with universal energies.

Parallel Reality: A concept that suggests the existence of multiple dimensions or versions of reality that can be accessed through shifts in consciousness or collective belief.

Quantum Physics: A branch of physics that studies the behavior of matter and energy at the molecular, atomic, nuclear, and even smaller microscopic levels, often drawing parallels to spiritual concepts of interconnectedness and consciousness.

Self-Discovery: The process of gaining deep insight into one's true nature, purpose, and spiritual essence, often through introspection, meditation, or life experiences.

Spiritual DNA: A metaphorical concept that represents an individual's unique spiritual essence, encoding information about past lives, spiritual purpose, and potential for growth. It is described as influencing one's experiences, lessons, and evolution across multiple lifetimes.

Spiritual evolution: The process of growth and development of consciousness across lifetimes, leading to higher states of awareness and connection with the Divine.

Subconscious mind: The part of the mind that operates below the level of conscious awareness, often associated with intuition, creativity, and access to spiritual wisdom.

Universal Life Force: The fundamental energy or consciousness that permeates all existence, often referred to in spiritual traditions as the source of all creation.

Vibratory frequency: The rate at which energy vibrates, often used metaphorically to describe levels of consciousness.

Bibliography

Adebayo, B. A., & Olisa, O. (2022). Health risk assessment of the consumption of geophagy clay from Mowe and Ikorodu markets, southwestern Nigeria. Goldschmidt2022 Abstracts.

Adebisi, S., Oyedele, A. A., & Adelakun, C. G. (2020). Appraising the efficacy of the National Economic Empowerment and Development Strategy (Needs) as policy intervention in Nigeria (2003-2007). International Journal of Economic Behavior, 9, 67–83.

Adiasto, K., Hooff, M. L. M., Beckers, D. G. J., & Geurts, S. (2023). The sound of stress recovery: an exploratory study of self-selected music listening after stress. BMC Psychology, 11.

Adjei, A., Amevinya, G., Quarpong, W., Tandoh, A., Aryeetey, R., Holdsworth, M., Agyemang, C., Zotor, F., Laar, M., Mensah, K., Addo, P., Laryea, D., Asiki, G., Sellen, D., Vandevijvere, S., & Laar, A. (2022). Availability of healthy and unhealthy foods in modern retail outlets located in selected districts of Greater Accra Region, Ghana. Frontiers in Public Health, 10.

Afghan, J. (2024). Death – An Inevitable Fact of Life: A Comparative Study of the Portrayal of Death in Emily Dickinson's and Jalaluddin Rumi's Selected Poems. International Journal of Linguistics and Translation Studies.

Afrin, F., & Islam, M. A. (2023). Robust Detection of Static and Moving Chipless RFID Tags Using Scalable Tree Boosting System. IEEE International WIE Conference on Electrical and Computer Engineering, 1–5.

Agrawal, V., Arjona, L. D., & Lemmens, R. (2001). E-performance: The path to rational exuberance. 31–31.

Ahmed, S., Altaf, N., Ejaz, M., Altaf, A., Amin, A., Janjua, K., Arif-Khan, Imran, I., & Khan, S. (2020). Variations in the frequencies of polymorphisms in the CYP2C9 gene in six major ethnicities of Pakistan. Scientific Reports, 10.

Amal, T., Banga, A., Bhatt, G. J., Faisal, U. H., Khalid, A., Rais, M. A., Najam, N., Surani, S. R., Nawaz, F. A., & Kashyap, R. (2024). Guiding principles for the conduct of the Violence Study of Healthcare Workers and System (ViSHWaS): Insights from a global survey. Journal of Global Health, 14.

Amato, A., Becci, A., Bollero, A., Cerrillo-Gonzalez, M. M., Cuesta-López, S., Ener, S., Dirba, I., Gutfleisch, O., Innocenzi, V., Montes, M., Sakkas, K., Sokolova, I., Vegliò, F., Villen-Guzman, M., Vicente-Barragan, E., Yakoumis, I., & Beolchini, F. (2023). Life Cycle Assessment of Rare Earth Elements-Free Permanent Magnet Alternatives: Sintered Ferrite and Mn–Al–C. ACS Sustainable Chemistry and Engineering, 11, 13374–13386.

Amicucci, G. L., & Fiamingo, F. (2017). Usage of RFId in safety applications. 2017 IEEE International Conference on Environment and Electrical Engineering and 2017 IEEE Industrial and Commercial Power Systems Europe (EEEIC / I&CPS Europe), 1–6.

Anderson, D. (2021). The Soul's Logical Life and Jungian Schisms. Psychological Perspectives, 64, 37–53.

Andreeva, O. (2019). Teaching text activity at the Russian lessons under the conditions of multicultural educational space. Pedagogy Issues of Theory and Practice.

Angels, N. L., & Elsherbeni, A. Z. (2024). Safety of Wireless Brain Implants: A Review. International Conference on Advances in Cybersecurity, 1–2.

Archer, J. E., Baird, C., Gardner, A., Rushton, A., & Heneghan, N. (2021). Evaluating measures of quality of life in adult scoliosis: a protocol for a systematic review and narrative synthesis. Systematic Reviews, 10.

Aunan, K., Orru, H., & Sjödin, H. (2024). Perspectives on connecting climate change and health. Scandinavian Journal of Public Health, 14034948241269748.

Baltezarević, B. V. (2023). The dark horizon of artificial intelligence and ChatGPT: A transhumanistic perspective. Knowledge - International Journal.

Barkam, S., Saraf, S., & Seal, S. (2013). Fabricated micro-nano devices for in vivo and in vitro biomedical applications. Wiley Interdisciplinary Reviews: Nanomedicine and Nanobiotechnology, 5(6), 544–568.

Bartl, A., Jung, S., Kullmann, P., Wenninger, S., Achenbach, J., Wolf, E., Rack, C., Lindeman, R., Botsch, M., & Latoschik, M. E. (2021). Self-Avatars in Virtual Reality: A Study Protocol for Investigating the Impact of the Deliberateness of Choice and the Context-Match. 2021 IEEE Conference on Virtual Reality and 3D User Interfaces Abstracts and Workshops (VRW), 565–566.

Bauer, K. (2007). Wired Patients: Implantable Microchips and Biosensors in Patient Care. Cambridge Quarterly of Healthcare Ethics, 16, 281–290.

Brown, R., Chevalier, G., & Hill, M. (2010). Pilot study on the effect of grounding on delayed-onset muscle soreness. Journal of Alternative and Complementary Medicine, 16(3), 265–273.

Bye, Z. L., Keshavarz, P., Lane, G., & Vatanparast, H. (2021). What Role Do Plant-Based Diets Play in Supporting the Optimal Health and Well-being of Canadians? A Scoping Review. Advances in Nutrition.

Chevalier, G. (2010). Changes in pulse rate, respiratory rate, blood oxygenation, perfusion index, skin conductance, and their variability induced during and after grounding human subjects for 40 minutes. Journal of Alternative and Complementary Medicine, 16(1), 1–7.

Chevalier, G., Mori, K., & Oschman, J. L. (2006). The effect of Earthing (grounding) on human physiology. European Biology and Bioelectromagnetics, 2(1), 600–621.

Chevalier, G., Sinatra, S. T., Oschman, J. L., & Delany, R. M. (In press). Grounding the human body reduces blood viscosity—a major factor in cardiovascular disease. Journal of Alternative and Complementary Medicine.

Chevalier, G., & Sinatra, S. (2011). Emotional stress, heart rate variability, grounding, and improved autonomic tone: clinical applications. Integrative Medicine: A Clinician's Journal, 10(3).

Chandrana, Dr. A. K. (2023). Correlating Language and Music for the Activation of Human Mind. The Creative Launcher.

Cortez, A. D., Bolnick, D. A., Nicholas, G., Bardill, J., & Colwell, C. (2021). An ethical crisis in ancient DNA research: Insights from the Chaco Canyon controversy as a case study. Journal of Social Archaeology, 21, 157–178.

Dakshit, S., Zakir, Md. F., & Mitra, A. (2017). Enhanced Protection of Implants from Potential Threats of Data Theft and Misuse. International Journal of Engineering Research and Technology, 6.

Das, S. (1997). Awakening the Buddha within: eight steps to enlightenment: Tibetan wisdom for the Western world.

Erdal, B., Tepe, Y., Çelik, S., Güçyetmez, B., Çiğdem, B., & Topaktas, S. (2021). The magic of frequencies - 432 Hz vs. 440 Hz: Do cheerful and sad music tuned to different frequencies cause different effects on human psychophysiology? A neuropsychology study on music and emotions. Journal of Human Sciences.

Ewen, B. M. (2020). The Connection Between the Body and the Environment: a Changing View. Science Education, 29, 1093–1096.

Feynman, R., Leighton, R., & Sands, M. (1963). The Feynman Lectures on Physics (Vol. 2). Addison-Wesley.

Holiday, D., Resnick, R., & Walker, J. (1993). Fundamentals of Physics (4th ed.). John Wiley & Sons.

Hong, E.-J., & Kangas, M. (2021). The Relationship between Beliefs about Emotions and Emotion Regulation: A Systematic Review. Behaviour Change, 39, 205–234.

Hung, S. (2012). Dharma Therapy A Therapeutic Intervention That Builds On The Buddhist Dharma With Mindfulness Practice As One Of Its Key Components.

Just, A. (1903). Return to Nature: The True Natural Method of Healing and Living and The True Salvation of the Soul. B. Lust.

Ku, C., Yang, Y., Park, Y., & Lee, J. (2013). Health benefits of blue-green algae: prevention of cardiovascular disease and nonalcoholic fatty liver disease. Journal of Medicinal Food, 16(2), 103–111.

Kumar, M., Abhayapala, T., & Samarasinghe, P. (2022). A Preliminary Investigation on Frequency Dependant Cues for Human Emotions. Acoustics.

Li, Y., & Lee, F. (2023). Influence of Music Psychology on the Consistency of Mood in Movies: A Review of Historical Works and Empirical Research. Herança.

Liang, G., Li, Y., Liao, D., Hu, H., Zhang, Y., & Xu, X. (2019). The Relationship between EEG and Depression under Induced Emotions Using VR Scenes. 2019 IEEE MTT-S International Microwave Biomedical Conference (IMBioC), 1, 1–4.

Liga, F., Cannavò, M., Papa, F., & Cuzzocrea, F. (2024). The relationship between Emotions, Beliefs, and Pro-Environmental Behaviors in Young Adults through the lens of Self-Determination Theory. International Journal of Emotional Education.

Michael, K., & Michael, M. (2013). The future prospects of embedded microchips in humans as unique identifiers: the risks versus the rewards. Media, Culture & Society, 35, 78–86.

Nurgaliyev, S., & Pradeep, A. (2023). Advancements in Brain Implants: Bio Sensors, IoT and Security Concerns. International Conference on Circuit, Power and Computing Technologies, 40–44.

Pebsworth, P., Seim, G. L., Huffman, M., Glahn, R., Tako, E., & Young, S. (2013). Soil Consumed by Chacma Baboons is Low in Bioavailable Iron and High in Clay. Journal of Chemical Ecology, 39, 447–449.

Perera, D. N., Hewavitharana, G. G., & Navaratne, S. (2021). Comprehensive Study on the Acrylamide Content of High Thermally Processed Foods. BioMed Research International, 2021.

Price, W. (1943). Nutrition and Physical Degeneration. Keats Publishing.

Qian, F., Zhou, S., & Zuo, L. (2020). Improving the off-resonance energy harvesting performance using dynamic magnetic preloading. Acta Mechanica Sinica, 36, 624–634.

Rossi, W. (1989). The Sex Life of the Foot and Shoe (Vol. 61). Wordsworth Editions.

Rudenko, S., Sobolievskyi, Y., & Zhang, C. (2021). Feng Shui Cosmology and Philosophy in Native Americans' Worldview. Philosophy and Cosmology.

Saad, M., Medeiros, R., & Mosini, A. (2017). Are We Ready for a True Biopsychosocial–Spiritual Model? The Many Meanings of "Spiritual." Medicines, 4.

Saha, S. (2015). Anagarika Dharmapala's Concept and Vision of Buddha and Buddhism as Enunciated in His Chicago Address, The World's Debt to Buddha: A Critical Enquiry and Assessment.

Saiz-Clar, E., Serrano, M., & Reales, J. (2021). Predicting emotions in music using the onset curve. Psychology of Music, 50, 1107–1120.

Salakka, I., Pitkäniemi, A., Pentikäinen, E., Mikkonen, K., Saari, P., Toiviainen, P., & Särkämö, T. (2021). What makes music memorable? Relationships between acoustic musical features and music-evoked emotions and memories in older adults. PLoS ONE, 16.

Shafi, A. (2020). Music as an Expression of Emotions. 68, 1298–1304.

Sherwani, R., Waheed, A., & Gulzar, S. (2023). Clean Energy as an Alternative: Identification of Factors Determining the Willingness of Slum Dwellers. Sir Syed University Research Journal of Engineering & Technology.

Shet, K., Tripathi, N., Varma, R., & Mishra, G. (2023). Unlocking Peak Performance: Chakra Balancing with Solfeggio Frequencies for University Level Working Professionals Enhancing Memory and Attention.

Silva, M. A. D., Espirito Santo, G. F., Manzolli, J., & Queiroz, M. (2021). A Psychoacoustic-Based Methodology for Sound Mass Music Analysis. Computer Music Modeling and Retrieval, 267–281.

Sokal, K., & Sokal, P. (2012). Earthing the human organism influences bioelectrical processes. Journal of Alternative and Complementary Medicine, 18(3), 229–234.

Spytska, L. (2024). Conscious, unconscious, and subconscious: The relationship between the three levels of human mental activity and their impact on life. Salud, Ciencia y Tecnología.

Stein, R. (2008). Is Modern Life Ravaging Our Immune Systems? Washington Post.

Vayenas, C., & Souentie, S. (2012). The Strong Force: From Quarks to Hadrons and Nuclei. 23–29.

Vaziri, M. (2012). Introduction to the Buddha's Key Spiritual and Philosophical Concepts. 3–12.

Vessia, F., & Muciaccia, N. (2018). ICT Implants and Brain-Computer Interfaces: Legal Issues in the EU Framework. Innovation Law & Policy eJournal.

White, G. (1929). The Finer Forces of Nature in Diagnosis and Therapy. Phillips Printing Company.

Wong, Y. H., Lee, G. C., & Sim, H. K. (2023). RFID and Facemask Detector Attendance Monitoring System. International Journal on Robotics Automation and Sciences.

Wuebben, D. (2021). Of robots and rhetoric: Nikola Tesla's telautomaton and the boundaries of scientific communication (1897–1900). Public Understanding of Science, 30, 484–492.

Book Review Request

Dear reader,

Thank you for purchasing this book! I would love to know your opinion. Writing a book review helps in understanding the readers and also impacts other readers' purchasing decisions. Your opinion matters. Please write a book review!

Your kindness is greatly appreciated!

About the Author

D an Desmarques is a renowned author with a remarkable track record in the literary world. With an impressive portfolio of 28 Amazon bestsellers, including eight #1 bestsellers, Dan is a respected figure in the industry. Drawing on his background as a college professor of academic and creative writing, as well as his experience as a seasoned business consultant, Dan brings a unique blend of expertise to his work. His profound insights and transformational content appeal to a wide audience, covering topics as diverse as personal growth, success, spirituality, and the deeper meaning of life. Through his writing, Dan empowers readers to break free from limitations, unlock their inner potential, and embark on a journey of self-discovery and transformation. In a competitive self-help market, Dan's exceptional talent and inspiring stories make him a standout author, motivating readers to engage with his books and embark on a path of personal growth and enlightenment.

Also Written by the Author

1. 66 Days to Change Your Life: 12 Steps to Effortlessly Remove Mental Blocks, Reprogram Your Brain and Become a Money Magnet

2. A New Way of Being: How to Rewire Your Brain and Take Control of Your Life

3. Abnormal: How to Train Yourself to Think Differently and Permanently Overcome Evil Thoughts

4. Alignment: The Process of Transmutation Within the Mechanics of Life

5. Audacity: How to Make Fast and Efficient Decisions in Any Situation

6. Beyond Self-Doubt: Unleashing Boundless Confidence for Extraordinary Living

7. Breaking Free from Samsara: Achieving Spiritual

Liberation and Inner Peace

Wrong

Any Problem

About the Publisher

This book was published by 22 Lions Publishing.

www.22Lions.com

www.ingramcontent.com/pod-product-compliance
Lightning Source LLC
Chambersburg PA
CBHW052328220526
45472CB00001B/326

* 9 7 9 8 3 4 6 4 1 0 6 6 9 *